DATE DUE

SEP 2 8 2017		
JAN 0 6 2020		
		PRINTED IN U.S.A.

Life of Fred®

Honey

Life of Fred®
Honey

Stanley F. Schmidt, Ph.D.

Polka Dot Publishing

ISBN: 978-1-937032-01-2

Library of Congress Catalog Number: 2011915651
Printed and bound in the United States of America

Polka Dot Publishing   Reno, Nevada

To order copies of books in the Life of Fred series,

visit our website PolkaDotPublishing.com

Questions or comments? Email the author at lifeoffred@yahoo.com

Fourth printing

Life of Fred: Honey was illustrated by the author with additional clip art furnished under license
from Nova Development Corporation, which holds the copyright to that art.

for Goodness' sake

or as J.S. Bach—who was
never noted for his plain
English—often expressed it:

Ad Majorem Dei Gloriam
(to the greater glory of God)

If you happen to spot an error that the author, the publisher, and the printer missed, please let us know with an email to: lifeoffred@yahoo.com

As a reward, we'll email back to you a list of all the corrections that readers have reported.

A Note Before We Begin
Life of Fred: Honey

Life gets so busy sometimes.

Wheaties

honey

Worcestershire   daughter

This was our breakfast table 41 years ago.

✶ You know that life is getting too complicated and distracting when you put the Worcestershire on your Wheaties.
✶ You know that you may not be gifted with foresight if you give your daughter an open jar of honey to play with.

What's the most important item on the table?

❀   ❀   ❀

Taking the long view in life is central to your ultimate happiness. Compare:
    ✓ My daughter called me yesterday to share some happy stories.
    ✓ The Worcestershire sauce bottle has never once bothered to call or even send me an email.
    ✓ The Wheaties box has spent the most recent 40 years of its dissolute life in a California landfill.

❀   ❀   ❀

Kids need two things in order to have a sunshine-filled rest of their lives. First, they need to be soaked in love from their parents and other concerned adults. Second, they need a real education.

real education = broad and deep

Broad: In the government schools they herd 25 students into a room, and someone talks to them about history for 50 minutes. A bell rings and the students head into another room, and someone talks to them about art for 50 minutes. A bell rings and they head into another room, and someone drills them on their math tables.

This is unnatural. There is an essential inner coherence among all the areas of learning. We are supposed to be teaching children—not subjects. In this book, the reader will learn what an apiarist is, what it means to pencil out a proposed business venture, how to make steel, why bees make their honeycombs in the shape of hexagons and not squares or octagons, and how you can tell terbium from copper. All these arise in Fred's everyday life.

And we even do a bit of math!

Deep: It's real simple. I know of no other math curriculum (Saxon, Singapore, Math-U-See, Teaching Textbooks, etc.) that contains more mathematics than the Life of Fred series.

HOW THIS BOOK IS ORGANIZED

Each chapter is about six pages. Have a paper and pencil handy before you sit down to read so you can do the Your Turn to Play at the end of each chapter.

Don't just read the questions and look at the answers. Your child won't learn as much taking that shortcut.

In Chapter 12, the reader will be asked to make Fred's Honey Cards. It will take five sheets of paper, scissors, and a pencil. The total cost will be around 6¢.

CALCULATORS?

Not now. There will be plenty of time later when you hit pre-algebra. Right now in arithmetic, our job is to learn the addition and multiplication facts by heart. That's where Fred's Honey Cards will come in handy.

Contents

Chapter One
Colors

When something new happens in your life, you often dream about it that night. That happened to Fred.

On Saturday, he got a goldfish and named it Fish. In the evening he watched it swim around in the tank while he talked to it about checkers, bicycles, dancing, fountain pens, and a zillion other topics.

It was now early Sunday morning. Fred was tucked into his sleeping bag under his desk.

He dreamed that Fish was flying around above his tank. Fish seemed to float in the air the way that he had floated in the water.

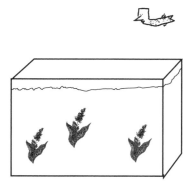

Fish seemed very happy. You could tell by the smile on his face. With silent fish-talk, he asked Fred if he would like to go swimming in the tank. Fred wanted to.

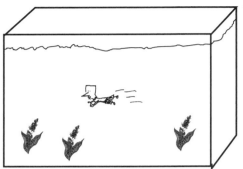

Each plant in the tank had six colors on it. All three plants had 18 different colors.

Fred could breathe underwater as easily as Fish could breathe the air.

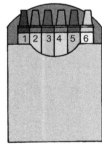

In his dream, every box of crayons was different.

One box

yellow, orange, red, purple, violet, blue

One box

green, black, magenta, cyan, pink, gray

One box

brown, lavender, maize, silver, copper, gold

Three different boxes of crayons could color all the plants. 3 × 6 = 18.

Fred colored each plant. This doesn't make much sense because every plant was already colored. But in a dream almost anything can happen.

Fred had no trouble breathing underwater in his dream, but he was starting to feel cold and wet.

He looked at his skin. It was turning blue. All shades of blue: azure mist, Alice blue, baby blue, periwinkle, powder blue, Cornflower blue, sky blue, aquamarine blue, turquoise blue, Ukrainian Azure, United Nations azure, cerulean, Bondi blue, steel blue, agate blue, indigo, slate blue, Dodger, royal blue, denim, Swedish azure, cobalt blue, Persian blue, lavender, International Klein blue, Ultramarine, navy blue, sapphire, midnight blue, Prussian blue, teal, Palatinate, Federal blue, Phthalo blue, and Air Force blue.

Fred awoke. He turned on the lamp and looked at the clock. It was much too early.

3:40 a.m.

He found out why he felt wet in his dream. He was wet. At first, he thought he had wet his sleeping bag. Then he realized that there was way too much water for that.

Then it hit him.

Oh my! Fred thought to himself. *There must be a plumbing leak. Or, maybe, it rained and the roof is leaking.*

Fred was wrong. The fish
tank had leaked during the
night.

He was sitting in five
gallons of water. Fish was in
zero gallons of water.

$$\begin{array}{r} 5 \\ -\ 5 \\ \hline 0 \end{array}$$

Fred climbed up onto his desk and took out
the three plants and put them on his desktop.
Those 18 colors would be fun to look at.

It was too early in the morning to call the
janitor. Fred didn't have any parents. He had
to clean everything up himself.

He took the tank and Fish down the
hallway past the nine vending machines (four on
the one side and five on the other), down the two
flights of stairs, and out into the cold February
night.

He said goodbye and put the tank and Fish
into the dumpster.

As he climbed the stairs, he counted them.

$$1 \quad 2 \quad 3 \quad 4 \quad 5 \quad 6 \quad 7 \quad 8 \quad 9 \quad 10 \quad 11 \ldots$$

These are the **natural numbers**, also
known as the counting numbers.

Fred dreamed of the day when he would be tall like Alexander. Alexander is about six feet tall.

Once Fred had seen Alexander go up the stairs three-at-a-time.

3   6   9   12   15   18   21   24   27

Please write out your answers. Don't just look at the questions and then look at the answers. Writing helps you to remember.

Your Turn to Play

1. A box of crayons contains 8 colors. How many crayons are there in 3 boxes?

2. The commutative law of multiplication says that 3 × 8 is the same as ___?___.

3. If a giant box of crayons had 77 colors in it, how many crayons would be in 3 giant boxes?

4. Sets are enclosed in braces. This is the set of whole numbers: {0, 1, 2, 3, . . . }.

If you add together two whole numbers, will your answer always be a whole number?

.......ANSWERS.......

1. $3 \times 8 = 24$

2. In algebra, we say that the commutative law of multiplication is $a \times b = b \times a$ (where a and b are any numbers).

   $3 \times 8 = 8 \times 3$ by the commutative law of multiplication.

3. $\overset{2}{7}7$ 3 times 7 is 21. Write down the 1 and
    $\underline{\times \ \ 3}$ carry the 2.
    231 3 times 7 is 21, plus 2, is 23.

4. The whole numbers are **closed under addition**: If you add two whole numbers, you will always get an answer that is a whole number.

   The whole numbers are not closed under subtraction.

   If you subtract 5 from 2, the answer can't be found in the set {0, 1, 2, 3, 4, 5, 6, 7, . . .}.

$$\begin{array}{r} 2 \\ -\ 5 \\ \hline ? \end{array}$$

Chapter Two
Life's Mysteries

Fred looked at his wet sleeping bag. He knew that if he just left it on the floor, it wouldn't dry.

The laundry room is on the first floor of the Math building where Fred lives. Fred could barely lift the wet sleeping bag. He wouldn't be able to carry it down the hallway and down two flights of stairs.

small essay

Problems

Much of life consists of solving problems. Fred has the problem of getting a heavy sleeping bag down to the first floor. Mothers of young children have problems. Owners of pizza stores have problems. Mothers of teenage children have problems.

Be very happy if the career you choose has lots of problems to solve! That will mean that you have a job that matters.

Would you want a job that consisted of just drawing small circles on a piece of paper all day long?

end of small essay

Fred wondered, *How can I get that heavy sleeping bag down to the ground level?*

He opened the window in his office and pushed the sleeping bag through it. *It would land on the sidewalk, and from there it would be easy to drag it into the laundry room.*

At least, that is what Fred thought.

He closed the window and raced down the stairs to the ground floor. If he were going at the speed of 3 stairs per second, in 9 seconds he would have descended 27 stairs.

The distance you go is equal to the rate you are traveling times the amount of time.

One of the most important formulas you will have in beginning algebra is $d = rt$. Distance equals rate times time.

There were several drawbacks to Fred's idea of throwing the sleeping bag out the window.

#1: It might have landed in the dirt. Then he would have to wash the sleeping bag.

#2: It might have hit someone.

#3: It might have landed on a car and broken the windshield.

When you get an idea, it is important to think of the drawbacks.

When Fred got down to the ground floor and headed outside, he couldn't find his sleeping bag. It wasn't 4 a.m. yet and everything was pretty dark. He had remembered to bring a flashlight just in case it was hard to find.

He looked all around.

✓ It couldn't have been carried away in the wind. It was too heavy, and there was no wind.

✓ It probably wasn't stolen. What thief would steal a wet sleeping bag at 3:55 a.m.?

✓ Fred was in the right spot—just below his window. He wasn't looking on the wrong side of the building.

Sometimes in life we encounter mysteries that we can't figure out. That's just part of living.

After 15 minutes* of looking, Fred headed back up to the third floor. In the hallway, he stopped at the janitor's closet and got a mop and a bucket so that he could clean up the water under his desk.

There were only three gallons of water under his desk. The fish tank originally

---

* That's a quarter of an hour.

had five gallons of water in it. Where did the other two gallons go? Another mystery. But this one was easy to figure out.

The sleeping bag had soaked up two gallons of the water.

$$\begin{array}{r} 5 \\ -\ 3 \\ \hline 2 \end{array}$$

    5   gallons in the fish tank
 – 3   gallons on the floor
    2   gallons in the sleeping bag

Fred knew that if he mopped up all three gallons of water into the pail, he would have trouble carrying it down the hallway to the restroom where he could empty it.

A gallon of water weighs about 8 pounds, so three gallons of water would weigh 24 pounds.

An obvious solution would be to mop up a gallon at a time. Fred could carry a gallon of water down the hallway.

He thought of a different solution. He mopped up a gallon of water and then poured it out of the window. He wasn't doing this to save energy. He had something else in mind.

He took his flashlight and ran down the two flights of stairs and outside. There was a big splash of water on the ground—but no sleeping bag. The mystery remained.

He carried the other two gallons of water down the hallway to the restroom, a gallon at a time and then returned the mop and pail to the janitor's closet.

Your Turn to Play

1. Fred started looking for his sleeping bag at 3:55 a.m. He looked for 15 minutes. What time is it now?

2. [Hard question. This may require two minutes of thought.]

The natural numbers are N = {1, 2, 3, 4, 5, 6, . . .}.

The **cardinal numbers** are the numbers that are used to count the number of members of any set.

For example, the cardinal number associated with the set {❀, ✂, ✪, ❀} is 4.

Your challenge is to show that the cardinal numbers are not the same as the natural numbers.

3. [Harder question. This may require five minutes of thought.]

The whole numbers are W = {0, 1, 2, 3, 4, 5, . . .}.

Show that the set of cardinal numbers is not the same as the whole numbers. You can do this by finding a set that is so large that the cardinal number associated with the set is not a whole number.

4. [The Guess-a-Function game.]

Here is a function.   6 → yes   33 → no   34 → yes

10 → yes   20 → yes   5 → no   8 → yes   45 → no

20,000 → yes   99,999 → no   98,629,624 → yes

Find the rule for this function.

. . . . ANSWERS . . . .

1. To go from 3:55 to 4:00 takes 5 minutes.
   To go from 4:00 to 4:10 takes 10 minutes. } 15 min.

2. If you just looked at the question and then turned the page and just looked at this answer, you lost something. Without making the effort,
   without giving it two minutes of thought,
   without really trying,
   you lost an opportunity to grow up,
   to strengthen your ability to concentrate,
   to develop character.*

The natural numbers {1, 2, 3, 4, 5, . . .} would not be enough to count the number of members of this set: { }. The cardinality of { } is 0, and 0 is not a natural number.

3. Each of the whole numbers {0, 1, 2, 3, 4, . . .} is finite (pronounced "FIY-night"). A finite number is a number that is not infinite ("IN-fin-nit"). It is not unboundedly large. The cardinality of {1, 2, 3, 4, . . .} or of the set {1, 10, 100, 1000, . . .} or of the set {2, 4, 6, 8, 10, 12, . . .} is not a whole number.

4. The rule is: *Assign each even number to "yes" and each odd number to "no."*

---

* *Character* is an old-fashioned word. It is a moral or ethical quality. It involves honesty and strength of will. A person of good character can face difficult situations without running away.

Chapter Three
Early Morning Hours

Normally, at a quarter after 4 on a Sunday morning, Fred might have headed back to bed to get some more sleep before dawn. But his sleeping bag had disappeared, and he didn't want to try to sleep on the hard floor of his office.

4:15 a.m.

If he were Kingie, he might spend the early morning hours painting or playing on his small grand piano.

If he were Joe, his not-too-bright student, he would just sit in front of a television set and eat potato chips.

But Fred is Fred. He put on his jogging clothes, grabbed his flashlight, and headed outside. Some people in Fred's situation might be unhappy that they couldn't have a couple more hours of sleep in a warm sleeping bag. But Fred saw it differently. He thought, *This is wonderful. I've got a couple extra hours to go jogging and explore new places.*

It's not the situation that counts as much as what goes on in your head.

Fred thought, *If I had enough time this morning, I could jog all over the country.*

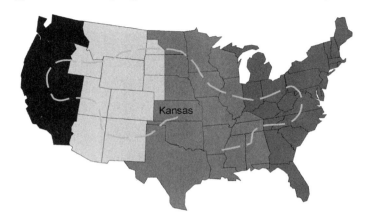

There are four time zones on this map. If it were 4:30 in the Central Time Zone (where Kansas is), then it would be 3:30 in the Mountain Time Zone and 2:30 in the Pacific Time Zone.

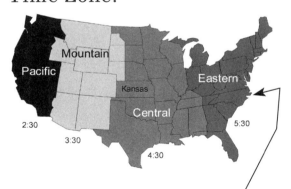

In the Eastern Time Zone it would be one hour later than the time in Kansas.

If you live in North Carolina and it's 7 a.m. and you call your friend on the Pacific coast, you are going to wake him up. It's only 4 a.m. there.

On the other hand, suppose you live in California, which is in the Pacific Time Zone. If it's 8 p.m. and you call your friend in North

Carolina, you may wake him up. It will be 11 p.m. there.

In the old days, before telephones were invented, you didn't have to worry about this. The letter that you mailed to your friend would never wake him up.

❀   ❀   ❀

Fred thought to himself, *I really don't have enough time this morning to jog all over the country. I've got about three hours. I run at the rate of six miles per hour (mph). That means I can cover about 18 miles.*

$$d = rt$$
distance equals rate times time

```
  6  mph
× 3  hours
─────
 18  miles
```

Which direction shall I head? I don't think I have ever headed southeast from KITTENS. That will be fun. I'll get to see some new stuff.

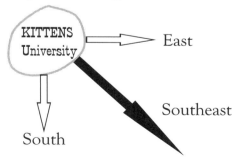

Fred didn't know what to expect when he headed southeast from KITTENS. It could be farmland. It could be an industrial area with lots of factories. It could be a commercial area with all kinds of stores. It could be a residential area with lots of homes.

To the southeast lay acres of homes.

As Fred ran through the neighborhoods, he found that he didn't need his flashlight because the street lights were on.*

Street lights make it easy to see where you are going. You won't accidentally run into a dinosaur.

Street lights make it hard to see the stars. Many kids growing up in the city have no idea what the night sky really looks like. Many have never seen the Milky Way. (They think that it is just a candy bar.)

Dinosaur drinking Sluice for breakfast. (Now you know why dinosaurs became extinct.)

If you go out into the countryside, you will get away from the street lights. On a clear night, you will see the Milky Way—a luminous white band stretching across the sky. There are so many stars in the band that only astronomers have

---

* When Fred had been in the countryside at night, it was **dark**.

been able to count them. Kids seeing the real night sky do so with their mouths open. Some have said, "I no longer want to be a ballerina or a baseball player. I want to be an astronomer."

Astronomers have counted the stars in the Milky Way. There are about a trillion stars in it. 1,000,000,000,000 stars. That's a million million stars.

Your Turn to Play

1. To multiply by 10, you just add a zero.

$26 \times 10 = 260$ or $777 \times 10 = 7{,}770$ or $50 \times 10 = 500$

To multiply by a 100, you add two zeros.

How many zeros do you add to multiply by a million? 1,000,000.

2. A trillion is 1 followed by how many zeros?

3. Astronomers have counted more than just the number of stars in the Milky Way. They have looked to the farthest stars. These are stars that are so far away that it would take millions of years for light to travel from them to our eyes.

Astronomers do much more than just count things, but counting is part of the fun. They have looked at the whole observable universe and counted the number of protons in it! There are 10^{79} protons. 10^{79} means $10 \times 10 \times 10 \times 10 \ldots \times 10 \times 10$—79 tens multiplied together.

Your question: What does 2^4 equal when you multiply it out?

. . . . ANSWERS . . . .

1. When you multiply something by a million, you add six zeros.

   $523 \times 1{,}000{,}000 \ = \ 523{,}000{,}000$

   $40 \times 1{,}000{,}000 \ = \ 40{,}000{,}000$

2. A trillion is 1,000,000,000,000, which is one followed by 12 zeros. This can be written as 10^{12}.

3. $2^{4} = 2 \times 2 \times 2 \times 2 = 16$

The whole observable universe contains 10^{79} protons. That's 10,000,000,000,000,000, 000,000,000,000,000,000,000,000,000,000, 000,000,000,000,000,000,000,000,000,000 protons. That's a finite number.

For most people (donut makers, dress makers, and paleobiologists), 10^{79} is a pretty big number. For mathematicians, the natural number 10,000,000,000,000,000, 000,000,000,000,000,000,000,000,000,000,000,000, 000,000,000,000,000 is actually pretty dinky. In the set of natural numbers {1, 2, 3, 4, . . .}, most members of that set are larger than 10^{79}. (10^{79} is read as ten to the seventy-ninth power.)

10^{79} protons. What's a **proton**? That's a question from chemistry. Chemistry is usually studied in about the eleventh grade, so this is the perfect time to give you a little preview of that subject.

Astronomers look at the really big things like the Milky Way. Chemists look at really small things like **atoms**.

Mangos, moose, and moms are made up of atoms. Everything you touch is made of atoms. They are really small. If a million atoms wacked you on the forehead, you wouldn't notice.

There are about a hundred different kinds of atoms. Many of them you probably have never heard of, such as terbium and dysprosium. But some may be familiar, such as iron, copper, calcium, aluminum, and carbon.

Here is a super simplified picture of an atom.

The black dots in the middle are protons. (The little gray dots flying around the protons are electrons. The white dots in the middle are neutrons.)

You can tell what kind of atom it is by the number of protons. Iron atoms always have 26 protons. Carbon atoms always have 6 protons.

If you want to see the whole list, look for **the periodic table of the elements.**

Chapter Four
Houses

F red ran through the neighborhoods. He looked at all the buildings. Some were apartment houses, and some were single family houses. All of them were bigger than the office that Fred had lived in for most of his life.

For a five-year-old, his office was the perfect size. Fred thought to himself, *When I grow up, I will probably need more space. My book collection will get too large.**

What Fred noticed was that houses were like people. They come in many varieties.

For people who like tall skinny windows

For people who don't like to spend a lot on upkeep

For people who like snug and secure

For people who like windows

For people who like stone castles

For people who like lots of garden

---

* At the age of five, Fred wasn't thinking about a wife and kids yet.

For people who like
a lot of rooms

For people who like
freeway access

For people who don't
demand privacy

Fred thought of the **function** where the domain would be the set of all people and the codomain would be the set of all houses. The function would be the rule that assigns to each person the one house that he or she likes the best.

For a rule to be a function, each member of the domain must be assigned to exactly one member of the codomain. That's the definition of a function.

As Fred jogged by the houses, he thought to himself, *I like the fact that there are so many different houses. If everyone were alike and wanted the same thing, this would be a very dull world.*

Fred giggled when he thought of what a phonebook would look like if everyone had the same name.

When Fred had jogged for an hour and a half, he knew that it was time to turn around and head back to his office. He knew that he wanted to run for only about three hours so he

would have plenty of time to get ready for
Sunday school. One and a half hours jogging
southeast plus one and a half hours jogging back
home equals three hours.

$$1\tfrac{1}{2} \quad + \quad 1\tfrac{1}{2}$$
$$= \quad 3$$

Fred had been heading southeast. When
he turned around to head back to his office, he
was heading northwest.

Fred had started jogging
at around 4:30. He had jogged
for an hour and a half toward the
southeast. When he turned around
it was about 6:00.

a compass

(4:30 to 5:30 is an hour. From 5:30 to 6:00
is a half hour.)

Fred felt so happy. He wanted to sing his MoRNiNG SoNG. If he were jogging in the countryside, he would have sung as he jogged.

But in the city, he might have awakened a lot of people who were still sleeping.

Please take out a piece of paper and write your answers down before you look at my answers on the next page. You will learn a lot more doing it that way.

Your Turn to Play

1. If Fred turned around at 6:00 and jogged northwest for an hour and a half, when would he arrive back at his office?

2. We have mentioned six of the eight points on a compass: N, E, S, W, SE, and NW. What are the other two points on a compass?

3. Fred was trying to invent a function where the domain was the set of all houses and the codomain was the set of all people.

His first try was the rule: Assign to each house the person who is living in the house now. Why would this *not* be a function?

4. His second try: Assign to each house the person who most recently touched the fireplace. Why would this *not* be a function?

5. His third try: Assign to each house the person who has most recently entered the house. Is this a function?

6. Is this a function? Assign every house to Millard Fillmore, the thirteenth president of the United States.

. . . . **ANSWERS** . . . .

1. From 6:00 to 7:00 is one hour.
    From 7:00 to 7:30 is a half hour. } 1 ½ hours

2. 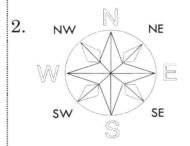   NE and SW.

3. If you want to determine if some rule is a function you have to be able to count up to one.

Every member of the domain must have *exactly one* assignment in the codomain.

   Does every house have exactly one person living in it right now? Of course not.

4. Suppose the rule is: Assign to each house the person who most recently touched the fireplace. My house would not be assigned to anyone. It doesn't have a fireplace.

5. I think this would be a function. Even vacant new houses had people walk inside of them. (For example, the carpenters that built the house.)

6. Every house is assigned to exactly one member of the codomain. It is a function. (It's called a **constant function** since every member of the domain is assigned to the same member of the codomain.)

Chapter Five
His Sleeping Bag

Fred decided to hum instead of sing. Then he wouldn't wake anyone up. Since the sun comes up in the morning, here is the song that Fred composed.

When Fred would become a man, his voice would not be as high as a five-year-old's. It might sound like this:

You can imagine what Fred's Evening Song would sound like as the sun went down. Instead of the notes going up the scale, they would be going down.

Despite the very little traffic at seven o'clock on a Sunday morning in February, Fred stopped humming when he came to intersections where he had to cross the street. When you are three feet tall, you have to be extra careful because drivers can't see you as easily.*

"Good morning, Fred."

Fred recognized the voice. It was Alexander. He and Betty were out for a morning jog and had caught up with Fred.

Fred responded, "Good morning, A and B. It's a lovely morning to be out running."

Betty said, "We've been running for 45 minutes. Next week, we hope to get up to an hour."

Fred, politely, didn't mention that he was doing a three-hour jog this morning.

small essay

Bragging

Telling people how wonderful you are is often not a good idea.

They tell you that they just got a new Rag-A-Fluffy doll, and you brag and announce that you have three new Rag-A-Fluffy dolls.

---

* Even when you get taller, you have got to watch out for cars.

Years ago, I, your author, taught math at college. One afternoon in one class I explained linear programing in a class for business majors. It was the last class of the day for one of my students. She was walking across the street to the campus parking lot. A car hit her.

When I learned that she was in the hospital, I sent her a get-well card. It was returned to me a couple of days later. The envelope was unopened. It was marked, "Patient expired."

If they start to tell you about the day they spent at Disneyland, you interrupt and tell them that you spent a week at Disneyland.

Do you think people are going to like you because you brag?

Being "better" does not make you more desirable as a friend.

end of small essay

Fred had to run hard to keep up with Alexander and Betty. Their legs were much longer.

As they ran together, Fred told them about how his goldfish had died. He told them about how his sleeping bag had gotten wet and how he had thrown it out the window because it was too heavy to carry down the stairs to the laundry.

Then he mentioned the big mystery: the sleeping bag had disappeared.

Alexander said that he once had a sock missing after he had done his laundry, but he had never lost a whole sleeping bag.*

Betty giggled. "Do you remember the time we lost a car? You drove me to King KITTENS and we spent 20 minutes looking for your car in

---

* This was the opposite of bragging. He made Fred feel special.

the parking lot? We couldn't remember where you had parked it."

Alexander grinned. He had forgotten that incident.

As they approached the Math building, Fred showed Alexander and Betty the big splash of water on the ground that he had made to mark the spot where the sleeping bag should have fallen.

Alexander and Betty looked at the water on the ground. They also couldn't figure out where Fred's sleeping bag had gone.

When they looked up at the trees, they couldn't see the sleeping bag. But the trees were evergreen and were hard to see through.

Betty had an idea. She climbed to the second story of the Math building and looked out the window.

"I see it!" she yelled.

There it was—Fred's little three-foot sleeping bag caught in the tree.

The mystery of where Fred's sleeping bag had gone was solved.

Some of our mysteries get solved, and some we wonder about all of our lives. That's true for all of us.

Your Turn to Play

1. For every step that Betty took, Fred had to take two steps.

   If Betty took 78 steps, how many would Fred have to take to keep up with her?

2. For every step that Alexander took, Fred had to take three.

   If Alexander took 69 steps, how many would Fred have to take to keep up with him?

3. Betty had climbed to the *second* story of the Math building. Is *second* a cardinal or an ordinal number?

4. $2 \frac{1}{2} + 2 \frac{1}{2} = ?$

5. Find a value of x that makes this true: $5 + x = 13$.

6. If you were facing north and you turned around, you would be facing south.

   If you were facing NW and you turned around, which way would you be facing?

. . . . ANSWERS . . . .

1. Fred takes twice as many steps as Betty.

$$\begin{array}{r} \overset{1}{7}8 \\ \times 2 \\ \hline 156 \end{array}$$

Fred took 156 steps.

2. Fred takes three times as many steps as Alexander.

$$\begin{array}{r} \overset{2}{6}9 \\ \times 3 \\ \hline 207 \end{array}$$ Fred took 207 steps.

3. *Second* is an ordinal number. Ordinal numbers describe the order things come in: first, second, third, fourth. . . .

4. 2 ½ + 2 ½ = 5

  +   =

5. If x is 8, then 5 + x = 13 is true.

6. You would be facing SE.

A Row of Practice. *Do the whole row before you look at the answers.*

69	207	75	95	786
+ 57	− 8	× 2	× 3	+ 469
126	199	150	285	1255

Chapter Six
Enough Rope

Fred tried shaking the tree, but trees that are two stories tall don't shake much. He had to think of some other way of getting his sleeping bag out of the tree.

From below, he couldn't even see the sleeping bag, so throwing a basketball at it to dislodge it wouldn't work.

He couldn't climb the tree. There were no low branches he could grab.

Fred had a fourth idea. He said to Alexander, "I could rent a hot-air balloon and float up to the sleeping bag and grab it."

"How much is your sleeping bag worth?" Alexander asked.

"Probably around $40," Fred said.

Alexander shook his head and said, "I bet it would cost a lot more than $40 to rent a balloon."

"I've got it!" Fred said. "I have a fifth idea. We could go up to the second story where Betty is, and you could throw me out of the window onto the tree. Then I could throw the sleeping bag down. And then. . . ."

Fred stopped talking. There were too many obvious flaws in his fifth idea.*

Fred ran up to his office and got some rope. He put on his cowboy hat to 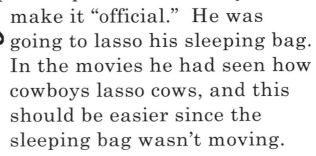 make it "official." He was going to lasso his sleeping bag. In the movies he had seen how cowboys lasso cows, and this should be easier since the sleeping bag wasn't moving.

He was ready to label his sixth idea as brilliant.

He ran down the stairs and stood under the tree. This wouldn't work. He couldn't see his sleeping bag.

He ran up the stairs to the second floor window where Betty was. Alexander followed him.

Fred explained, "I'ma gonna lasso that there sleeping bag and haul her in." He was trying to sound like a cowboy.

Alexander smiled and told Fred in a fake cowboy accent, "Listen, pardner, that there rope

---

* For example, what if Alexander missed the tree and Fred hit the ground instead? Or after Fred threw the bag out of the tree, how would Fred get out of the tree?

that you are aholding is a might tad too short
for the job. I reckon you'll need a longer rope."

Fred looked at his rope. He dropped the
cowboy talk and said simply, "Oops." Fred's
little six-foot rope wouldn't be long enough.

He ran down the stairs and paced off the
distance between the building and the tree. It
was 78 little Fred steps, which
equaled 78 feet.

$$78 > 6$$ (Seventy-eight is
greater than six.)

Fred ran up to the third floor hallway and
looked in the janitor's closet. Right behind the
four-gallon bottle of blue fountain pen ink, Fred
found exactly what he was looking for.

With the marks on the rope
at every yard, it would be easy to measure how
much rope Fred needed.

There are 3 feet in a yard. If he needed 12
feet, for example, he would divide by 3.

$$3\overline{)12}^{\,4}$$

Twelve feet would be the same as 4 yards.
This is called **division**. We divided 3 into 12
and got an answer of 4.

You could also say 12 divided by 3 is 4. That could be written as $12 \div 3 = 4$.

Since division is so important, we have a third way you can write it. You can express 12 divided by 3 as a fraction: $\dfrac{12}{3}$

$3\overline{)12}$ is the same as $12 \div 3$ is the same as $\dfrac{12}{3}$

If 3 hungry guys divide a dozen eggs into three equal groups, each guy will get 4 eggs.

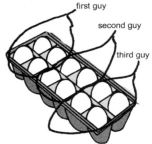

$12 \div 3 = 4.$

Officially, when you do division $3\overline{)12}^{\,4}$ there are two more steps.

First, after you put the 4 on top, you multiply the 4 times the 3 and put your answer under the 12.

$$3\overline{)12}^{\,4}$$
$$12$$

Second, you subtract. Since there is an answer of 0, that shows that 3 went into 12 evenly.

Remainder is zero.

Your Turn to Play

1. If three guys wanted to divide 13 eggs into three equal groups, there would be a problem. Divide 3 into 13 and show that there would be a remainder of 1.

2. Complete the next three terms in this sequence:

   4, 8, 12, 16, _?_, _?_, _?_.

3. If you were facing SW and you turned around, which way would you be facing?

4. $3\frac{1}{2} + 3\frac{1}{2} = ?$

5. Invent a function where the first set (the domain) is the set of all the people in Kansas and the second set (the codomain) is the set of all natural numbers ({1, 2, 3, 4, 5, 6, . . .}).

.......ANSWERS.......

1.
$$\begin{array}{r} 4 \\ 3\overline{)13} \\ 12 \\ \hline 1 \end{array}$$

Sometimes the remainder is written up on the line with the quotient like this ⇨

$$\begin{array}{r} 4\ \text{R}\ 1 \\ 3\overline{)13} \\ 12 \\ \hline 1 \end{array}$$

2. The sequence would be 4, 8, 12, 16, 20, 24, 28.

3. The opposite direction to SW is NE.

4. 3 ½ + 3 ½ = 7

$$\begin{array}{r} 3\ \ ½ \\ +\ 3\ \ ½ \\ \hline 6+1\ =\ 7 \end{array}$$

5. There are many possible answers. Your answer may be different than mine.

✓ My first answer: Assign everybody to the number 21. Then each member of the domain has exactly one image. It is a function. This function is called the constant function.

✓ My second answer: Assign everyone who has ever touched the cover of a copy of *Life of Fred: Honey* to 100. Assign everyone else to 96,309,230.

✓ My third answer: Assign everyone who has deciduous (baby) teeth to 4 and everyone else to 7.

I bet your answer was different than any of my answers. This is fun. I want to do some more.

✓ Assign the oldest person in Kansas to 1. The second oldest to 2. And so on.

✓ Assign all those who have ever eaten popcorn to 88 and the rest to 22.

Chapter Seven
Feet into Yards

There was 78 feet between the tree and the Math building. Fred needed to convert that to yards so that he could cut off the right amount of rope.

He needed to divide 3 into 78.

Here are all the steps . . .

$$\begin{array}{r} 2 \\ 3\overline{)78} \end{array}$$

> 3 doesn't go evenly into 7, but it does go evenly into 6. It goes into 6 two times.

$$\begin{array}{r} 2 \\ 3\overline{)78} \\ 6 \end{array}$$

> 2 times 3 is 6.

$$\begin{array}{r} 2 \\ 3\overline{)78} \\ \underline{6} \\ 1 \end{array}$$

> Subtract.

$$\begin{array}{r} 2 \\ 3\overline{)78} \\ \underline{6} \\ 18 \end{array}$$

> Bring down the 8.

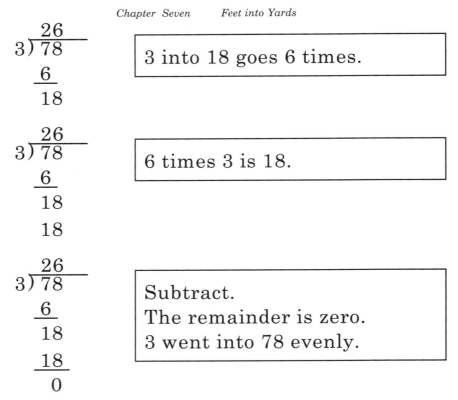

```
      26
   3) 78
      6
      18
```
3 into 18 goes 6 times.

```
      26
   3) 78
      6
      18
      18
```
6 times 3 is 18.

```
      26
   3) 78
      6
      18
      18
       0
```
Subtract.
The remainder is zero.
3 went into 78 evenly.

Fred had converted 78 feet into 26 yards.

small essay
The First Time

I remember the first time I took my first steps on my own. It was soooooooooo complicated. I had to decide which foot to start with. (Dancing a foxtrot the male starts with his left foot and the female starts with her right foot.)

Then I had to lift up that foot and move it forward in the air. Then I had to remember to put it down before I picked up the next foot. If I didn't I would fall right on my diaper.

Then after the left foot, I had to pick a different foot. If I did the left foot two times in a row, I started to do the splits.

Then I had to do the whole process over again with my other foot, which in this case I figured out was my right foot—up, forward, down. Forward, up, down didn't work.

Then, if I hadn't arrived at where I was going, I had to repeat the whole thing all over again.

Meanwhile, my mom is shouting at me, "Come on, Stan! I know you can do it!"

That really broke my concentration.

Learning to do division was almost as traumatic. I had to learn **divide, multiply, subtract, bring down**. And if I didn't arrive at where I was going, I had to repeat the whole thing over again: **divide, multiply, subtract, bring down**.

And if mom was cooking a delicious pizza for lunch, that really broke my concentration. How was I ever going to remember **divide, multiply, subtract, bring down?**

end of small essay

There is a giant example on the next page. Take a second and really study it. Go through each step. Instead of left—right—left—right, it will be **divide, multiply, subtract, bring down** followed by **divide, multiply, subtract, bring down**. It took you a week to learn to walk. Please give this three minutes.

$$
\begin{array}{r}
18794 \\
3\overline{)56382} \\
\underline{3} \\
26 \\
\underline{24} \\
23 \\
\underline{21} \\
28 \\
\underline{27} \\
12 \\
\underline{12} \\
0
\end{array}
$$

divide, multiply, subtract, bring down
divide, multiply, subtract, bring down
divide, multiply, subtract, bring down
divide, multiply, subtract, bring down
etc.

This example shows that 56,382 feet is the same as 18,794 yards.

Fred ran up to the third floor. In the janitor's closet he cut off 26 yards of rope.

He ran back to his office and put on some cowboy boots. Now he was ready to be a real cowboy.

He started to run down the stairs and found that it was hard to run with cowboy boots on. He tripped and fell down the stairs. If he had been 65 years old instead of 5, he might have broken a hip. (Hip fractures are much more common in older cowboys.) Fred found his hat and put it back on.

He walked to the second floor window where Alexander and Betty were waiting. He couldn't see the tree and the sleeping bag. He was too short.

Your Turn to Play

1. Convert 82,614 feet into yards.

A Note

♪#1: You have reached a milestone in your math education: one single harder problem in the Your Turn to Play.

When you were younger, you couldn't handle one big problem. You always received a bunch of little problems.

. . . . ANSWER . . . .

1.

$$
\begin{array}{r}
27538 \\
3{\overline{\smash{\big)}\,82614}} \\
\underline{6} \\
22 \\
\underline{21} \\
16 \\
\underline{15} \\
11 \\
\underline{9} \\
24 \\
\underline{24} \\
0 \\
\end{array}
$$

The remainder is zero. Three divided evenly into 82,614.

82,614 feet = 27,538 yards

A Second Note

♪#2: Some may ask, "Why do we gotta learn this division stuff? If I just punch 82614 into my calculator and then hit ÷ and then hit 3 and then hit =, out will pop the answer. It's so hard to remember divide, multiply, subtract, bring down."

I have three answers to that question.

<u>First Answer</u>: What career goals do you have? If you just plan to be a cashier at some fast food place, you won't need to learn

$$\begin{array}{r} 6 \text{ R } 2 \\ 3\overline{)20} \\ \underline{18} \\ 2 \end{array}$$

The little buttons on your cash register may even have little pictures on them, so you won't even have to learn to read!

Welcome to minimum wage for the rest of your life.

<u>Second Answer</u>: After you've done divide, multiply, subtract, bring down for a while, you'll be able to do $3\overline{)42}$ in your head. If you are in the middle of some real estate negotiations and 42 feet needs to be converted to yards, you will look pretty dorky having to haul out your calculator and punch in 42 ÷ 3 = when everyone else at the table knows it's 14 yards.

<u>Third Answer</u>: In *Life of Fred: Beginning Algebra*, we are going to divide polynomials. Here is where we really use divide, multiply, subtract, bring down.

$$\begin{array}{r} 4x \ + \ \ 2 + 3/(x+4) \\ x+4\overline{)\ 4x^2 + 18x + 11} \\ \underline{4x^2 + 16x} \\ 2x + 11 \\ \underline{2x + \ 8} \\ 3 \end{array}$$

Here is where no calculator will help.

Chapter Eight
Short Cowboy

Betty picked Fred up so that he could be tall enough to see the tree and the sleeping bag.

Fred twirled his lasso, gave a little cowboy yell, and threw the rope. He forgot to hold onto the other end of the rope. The rope landed in the puddle.

Fred took off his cowboy boots and ran down the stairs in his bare feet. He didn't want to fall again. He picked up the wet rope and headed back to the second floor.

Alexander picked him up this time. Fred's hat fell off. Betty held the other end of the rope. Fred twirled his lasso, gave another little cowboy yell, and threw the rope.

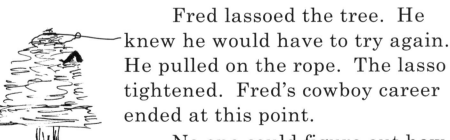 Fred lassoed the tree. He knew he would have to try again. He pulled on the rope. The lasso tightened. Fred's cowboy career ended at this point.

No one could figure out how to get the rope loose.

Fred picked up his cowboy boots and his hat. Betty tossed her end of the rope out the window. It was getting late. Fred needed to get ready for Sunday school. Alexander and Betty, for church.

Alexander headed to his apartment to shower and shave. Betty headed to her apartment to shower.

Fred headed to his office to put away his cowboy hat and boots. He headed down the hallway to wash off the dust from three hours of jogging. Kids don't sweat the same way adults do. And they don't shave.

He carried his towel in one hand and found a very handy way to carry his toothbrush and toothpaste. That left his other hand free to open doors.

Most people can't do this.

He stopped in front of one of the vending machines.

Bob's Better Breakfast

Bacon, Eggs, Pancakes, Sausage, Milk, Orange Slices, English Muffin

Insert Four Quarters

food comes out here

Four quarters, Fred thought to himself. *That's pretty cheap for a big breakfast.* Fred wasn't very hungry, but it started him thinking about his arithmetic class that he would be teaching tomorrow morning.

Four quarters buys one of 𝒷ℴ𝒷'𝓈 𝒷ℯ𝓉𝓉ℯ𝓇 𝒷𝓇ℯ𝒶𝓀𝒻𝒶𝓈𝓉𝓈. *How many of* 𝒷ℴ𝒷'𝓈 𝒷ℯ𝓉𝓉ℯ𝓇 𝒷𝓇ℯ𝒶𝓀𝒻𝒶𝓈𝓉𝓈 *could you buy with* 36 *quarters?*

For Fred, it seemed easy. All you would have to do is divide 4 into 36.

$$4\overline{)36}\ \ \ ^{9}$$

Then he realized that his students couldn't do that. They hadn't learned their four times table yet. They knew their three times table:

$$
\begin{array}{ccccccc}
3 & 4 & 5 & 6 & 7 & 8 & 9 \\
\times 3 & \times 3 & \times 3 & \times 3 & \times 3 & \times 3 & \times 3 \\
\hline
9 & 12 & 15 & 18 & 21 & 24 & 27
\end{array}
$$

Read these aloud!

Seven facts.

But he wondered how he could teach the six facts associated with the four times table:

$$
\begin{array}{cccccc}
4 & 5 & 6 & 7 & 8 & 9 \\
\times 4 & \times 4 & \times 4 & \times 4 & \times 4 & \times 4 \\
\hline
16 & 20 & 24 & 28 & 32 & 36
\end{array}
$$

Read these aloud!

Once his students learned their four times table, seeing that $4\overline{)36}\ \ ^{9}$ would be duck soup, a cinch, easy, no sweat, a walk in the park, effortless.

Even working with 37 quarters wouldn't be
hard.

$$\begin{array}{r} 9\ \text{R}\ 1 \\ 4\overline{)37} \\ \underline{36} \\ 1 \end{array}$$

Your Turn to Play

1. Without looking back, there's a good chance you
already know one of the four times table already.

    $4 \times 9 = ?$

2. Which of the four times table does this illustrate?

3. There are seven facts for the three times table.

There are six facts for the four times table.

There will be five facts for the five times table.
If I were to ask you how many facts for the nine times
table, you might just count

7 facts	3 times table
6 facts	4 times table
5 facts	5 times table
4 facts	6 times table
3 facts	7 times table
2 facts	8 times table
1 fact	9 times table.

There's a shorter way than doing all that counting.
Mathematicians are always looking for easier ways to
do things.

Look at each line. Do you see the pattern?

. . . . **ANSWERS** . . . .

1. We mentioned it four times before the start of this Your Turn to Play. $4 \times 9 = 36$

2. $4 \times 4 = 16$ Count them if you like.
Four rows and four columns make 16 ✳'s.

✳ ✳ ✳ ✳
✳ ✳ ✳ ✳
✳ ✳ ✳ ✳
✳ ✳ ✳ ✳

3. The first line mentions 7 and 3.
The second line mentions 6 and 4.
The next line mentions 5 and 5.

   For those who know their addition table, a little voice might say, "*Pow! Wham! Wow!* They all add to 10."

   So what goes with the 9 times table? Easy. 1 fact.

   Those that still count on their fingers or are married to a calculator, might miss the *Pow! Wham! Wow!*

   Learning the tables can enable you to see things that you might otherwise miss.

Chapter Nine
Math Easier Than English

Fred stood there in front of the *Bob's Better Breakfasts* vending machine staring at the sign that said Insert Four Quarters and thinking about teaching the four times table.

$$
\begin{array}{cccccc}
4 & 5 & 6 & 7 & 8 & 9 \\
\times 4 & \times 4 & \times 4 & \times 4 & \times 4 & \times 4 \\
\hline
16 & 20 & 24 & 28 & 32 & 36
\end{array}
$$

Read these aloud!

He thought to himself, *English teachers have it a lot rougher than we math teachers. They have to start off their youngest students with 26 new facts:* ABCDEFGHIJKLMNOPQRSTUVWXYZ.

They have to learn not only that T *is pronounced tea, but they have to memorize the order.*

Then they have to teach 26 more facts for the lowercase letters: abcdefghijklmnopqrstuvw xyz.

And they have to explain that lowercase a can be written either as a *or as* a.

Twenty-six uppercase letters. Twenty-six lowercase letters. Twenty-six pronunciations. And why you can write either a or *a*.

Please check to make sure I did the addition correctly.

$$\begin{array}{r} 26 \\ 26 \\ 26 \\ +\ 1 \\ \hline 79 \end{array}$$

Fred knew that English teachers had to teach a lot more than math teachers: 79 facts for kindergarten kids.

In Greece they have it a little easier. The Greek alphabet has only 24 letters. Every Greek kid has to learn these lowercase letters: αβγδεζηθικλμνξοπρστυφχψω.

Just as we have to learn that U is pronounced you, they have to learn that α is pronounced alpha.

Fred headed off to the restroom in the hallway as he added up all the multiplication facts.

8	facts for 2 times table
7	facts for 3 times table
6	facts for 4 times table
5	facts for 5 times table
4	facts for 6 times table
3	facts for 7 times table
2	facts for 8 times table
+ 1	fact for 9 times table

Please check to make sure I did the addition correctly.

36 multiplication facts

Fred thought to himself, *Wow. Only 36 multiplication facts and kids have several years to learn them. That beats 79 English facts to learn in kindergarten.*

When Fred got to the restroom, he set down his towel, toothbrush, and toothpaste. He was still trying to think how to teach the multiplication facts to his arithmetic class. *When I taught the addition facts, they were easier to illustrate. When I taught that 7 + 7 = 14, I could point out that there were 14 days in a fortnight.*

But how can I teach that 4 × 7 = 28? There are seven days in a week, but months usually have more than four weeks in them.

Maybe I could invent a tic-tac-toe game with four squares instead of three. Then my students could see that 4 × 4 = 16.

There are four cups in a quart, and there are four quarts in a gallon, so it will be easy to think of uses for the four times table. But none of these are as visual as the addition facts.

Fred picked up his toothbrush and toothpaste. *What if I squeeze the toothpaste four times as hard as I normally do? Will that help teach the four times table?* That didn't work. All that showed was that it was time to wash his head.

Fred
decorates
himself

Fred thought about the foreign language teachers at KITTENS. They have to teach their students about a thousand words. Even then, students would have difficulty reading a regular novel such as Graham Greene's *The Heart of the Matter* in German. In the first sentence: "Wilson . . . presste seine kahlen, rosigen Knie fest gegen das eiserne Gitter."*

And then the teacher has to explain that *presste* is the past tense for *pressen*. (Just as *ate* is the past tense for *eat* in English.)

Fred felt so good being a math teacher. Only six facts for the four times table.

$$
\begin{array}{cccccc}
4 & 5 & 6 & 7 & 8 & 9 \\
\times 4 & \times 4 & \times 4 & \times 4 & \times 4 & \times 4 \\
\hline
16 & 20 & 24 & 28 & 32 & 36
\end{array}
$$

Read these aloud!

---

* Wilson . . . pressed his bald pink knees against the iron railing.

Fred thought that if he just repeated them several times in the classroom, they would start to sink in. Or he could make it into a bit of a game by asking the class the multiplication questions, such as "Four times six?" and have the class respond with the correct answer.

At first, only a couple of students might say the answer aloud, but after a while more and more would join in.

(Parents who really want their kids to learn their math facts could spend two minutes in every car trip doing what Fred did in his class. Just two minutes. It might make a big difference.)

Your Turn to Play

1. What are the next four terms in the sequence
   3, 6, 9, 12, 15, _?_, _?_, _?_, _?_
2. $567 \times 100 = ?$
3. Name a cardinal number that is not a natural number.
4. If a tube of cerulean blue oil paint costs $7, how much would 3 tubes cost?
5. The Guess-a-Function game. Find the rule for this function.   six → 3   seven → 5   horse → 5   rat → 3
   million → 7   zipper → 6   pizza → 5   everything → 10
   nothing → 7   Fred → 4   Joe → 3   Alexander → 9
   eight → 5   I → 1   eye → 3   poem → 4   play → 4
   sedimentary → 11   rock → 4   me → 2   you → 3

........**ANSWERS** ....

1.   3, 6, 9, 12, 15, 18 , 21 , 24 , 27

2. $567 \times 100 = 56{,}700$. To multiply by a 100, just add two zeros.

3. The cardinal numbers count the number of members of a set. For example, the cardinal number associated with {Washington, Jefferson, Taft} is 3.

   The natural numbers are {1, 2, 3, 4, 5, 6, . . .}. The cardinal number associated with { } is 0. Zero is not a natural number.

4. If one tube cost $7, three tubes would cost $21.

5. The rule is: *assign the word to the number of letters in the word.*

A Row of Practice. *Do the whole row before you look at the answers.*

				283
49	553	48	639	338
+ 75	− 28	× 2	× 3	+ 778
124	525	96	1917	1399

Chapter Ten
Carrie

Fred washed the toothpaste off of his head, brushed his teeth, and headed back to his office. He changed out of his jogging clothes. When he taught his math classes, he wore a bow tie. He thought that made him look more like a teacher.

With a polka dot bow tie

But for Sunday school, he was just a student. It was fun, once a week, to be a student instead of the teacher.

8:50 a.m.

It was ten minutes to 9. Sunday school started at 9, so he had plenty of time to get there. It was only a three-minute walk from his office to the chapel on the KITTENS campus.

He got there at 8:53, seven minutes before class started.

"Good morning, Fred," Carrie said. She greeted each of her students by name.

"Hi," Fred responded. "I'm

Carrie

looking forward to your class. It's always a lot of fun."

She smiled, "The same is true of your class." During the week, Carrie studies calculus in Fred's class.

small essay

A Balance in Life

On Sunday, Carrie teaches Fred. On Monday, Fred teaches Carrie.

When you were a baby, your parents fed you and washed your clothes. When your parents get really old, you may get the chance to love them in the way that they loved you.

end of small essay

Carrie taught the Sunday school class for the five-year-olds. There were four tables with seven chairs at each table. (28!) Fred liked the chairs and tables because they were small. That meant that he only needed to sit on one phonebook to be tall enough to work at the table.

Without phonebook

Sitting on phonebook

By nine o'clock each chair had a five-year-old, and Carrie began, "Today, we are going to study something really sweet. We are going to study about honey. Honey is mentioned 61 times in the Bible."

Three of the kids were already thinking about snack time. One thought about having some cereal with honey on it. Another imagined graham crackers with honey on them. The third pictured a big bowl of honey with a spoon.

Fred imagined dividing 61 by 3:

$$\begin{array}{r} 20\ \text{R}\ 1 \\ 3)\overline{61} \\ \underline{6} \\ 01 \\ \underline{0} \\ 1 \end{array}$$

In that division, Fred had to multiply zero times three. The one times table is easy.

$$1 \times 1 = 1$$
$$1 \times 2 = 2$$
$$1 \times 3 = 3$$
$$1 \times 4 = 4$$
$$1 \times 5 = 5$$
$$1 \times 6 = 6$$
$$1 \times 7 = 7$$
$$1 \times 8 = 8$$
$$1 \times 9 = 9$$

But the zero times table is even easier.

$$0 \times 1 = 0$$
$$0 \times 2 = 0$$
$$0 \times 3 = 0$$
$$0 \times 4 = 0$$
$$0 \times 5 = 0$$
$$0 \times 6 = 0$$
$$0 \times 7 = 0$$
$$0 \times 8 = 0$$
$$0 \times 9 = 0$$

So $342,983,884,503,311,032 \times 0 = 0$.

Fred imagined teaching Sunday school. (He knew that he would probably be doing that after he grew up.)

He would teach God's multiplication:

$$\left(\begin{array}{c}\text{ALL OUR MISTAKES}\\ \text{ALL OUR SINS}\\ \text{ALL OUR EVIL DEEDS}\end{array}\right) \times 0 = 0 \quad \text{all gone!}$$

God's forgiveness

Carrie asked her class, "Where does honey come from?" These were five-year-olds, so she started off with an easy question.

Kelly raised her hand and said, "In our kitchen."

Chris said, "But we buy it at the store and take it home."

Terry said, "Bees make it and somehow deliver it to the store."

Bee Trucking

How Terry thought honey gets to the store

Lee said, "Bees can't drive trucks."

Dale said, "Bees chew up flowers and stuff the honey in the closets of their homes. That's what my brother told me."

Carrie smiled. It looked like she had a lot of explaining to do. "And what do we call the home that bees live in?"

Kelly raised her hand and said, "It's called an At Home, since if you be/bee home, you're at home."

Chris: A bee residence.

Terry: Home Sweet Home, since there's honey there.

Carrie explained, "It's called a hive."

Your Turn to Play

1. Carrie held up a honeycomb. "This is what Dale's brother called the closets where the bees store their honey."

part of a honeycomb

Lee said, "Those look like stop signs."

Was Lee right?

2. A question for artists. Bees make their honeycombs out of hexagons (6 sides). Could bees make a honeycomb using octagons (8 sides)? Draw your answer.

3. A hexagon has six sides. What do you call a shape with four equal sides?

. . . . ANSWERS . . . .

1. Stop signs have eight sides. Bees make honeycombs with six sides.

2. Can octagons (8 sides) fit together nicely like hexagons (6 sides) do?

It doesn't seem to work.

No matter how you arrange them, they don't fit together nicely. There is wasted space between the octagons.

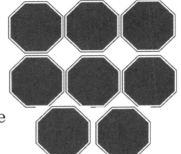

3. A four-sided figure with all four sides equal in length is called a rhombus (ROM-bus). The *h* is silent. If we also require that all the angles are equal, then we get a square.

rhombus   square

Advanced Honeycomb Question #1: Why don't bees use squares instead of hexagons?

Answer: Because squares could squish into rhombuses and all the honey would run out.

Advanced Honeycomb Question #2: Why don't bees use triangles? Triangles won't squish like squares.

Answer: The walls of a honeycomb have a thickness. It takes time and materials to build a wall. If you want to wall in a particular area, circles take the least wall material (perimeter). But there would be wasted space. (See the answer to question 2 in the Your Turn to Play.) Triangles take the most wall material to enclose a given area.

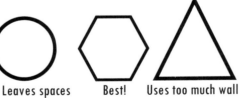

Leaves spaces   Best!   Uses too much wall

Chapter Eleven
Honey Cards

It was time for arts and crafts. Carrie knew that long lectures didn't work very well with five-year-olds.

Carrie took 28 sheets of paper. There were 4 tables. Each table got 7 sheets.

$$4\overline{)28}$$
$$\underline{28}$$
$$0$$

"We are going to make honey cards," she told the class.

Kelly raised her hand and said, "I don't know how to do that."

Carrie said, "I'll show you how. We are going to fold the paper to make it into eight rectangles."

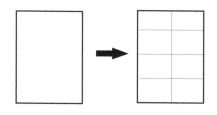

Terry said, "I don't get it."

Chris was tasting his paper to see if it tasted like honey.

"Now watch me," Carrie said, "and see how I fold it."

She held up a piece of paper so that everyone could see.

Terry said, "This is too hard."

Carrie folded it once.

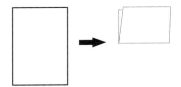

And then a second time.

Then she opened it up
and folded it once in the
other direction.

When she opened the paper
up, there were eight rectangles.

Chris asked for a new piece of paper. His
was all wet.

Terry had not paid attention.
Carrie gave Terry a new piece of
paper and asked Lee to help her.

Terry's paper

Fred noticed that the unfolded paper was
one rectangle—an odd number. But after you
started folding, there were always an even
number of rectangles—even on Terry's paper.

Carrie handed out plastic scissors to each child.

Fred thought of scissors that he had back in his office. They had been issued to all the teachers at KITTENS.

sharp steel

Time Out!

Do you know how to make steel? Here's how it's done.
First, you start with some iron.

How to Get Iron

Not this kind of iron!

You dig iron ore out of the ground. Iron ore has iron combined with oxygen ("rusted iron") or with sulfur ("fool's gold").

How to Make Steel

You smelt the iron ore.

What's Smelting?

I don't have room here to explain. Please see the next page.

What's Smelting?

Take the iron ore and remove the oxygen and the sulfur and add a little carbon. That's called smelting.

You do this all in a furnace that is many times hotter than your kitchen stove.

How Much Carbon?

You can add up to two percent (2%) carbon. That means 2 parts of carbon and 98 parts of iron.

Steel is much harder than iron.

A ton is equal to 2,000 pounds. Each year about 1,300,000,000 tons of steel are produced. (In words, one billion, three hundred million.)

Steel is used in making scissors for adults, tools, cars, ships, bridges, and pizza ovens.

Carrie told her students to carefully cut their paper into eight little rectangles.

Kelly raised her hand and asked, "Where do we cut?"

Carrie explained that they were to cut along the fold lines.

Fred was not used to using plastic scissors with blunt ends. The steel scissors back in his office worked much better.

But for five-year-olds, plastic scissors are much safer.

Your Turn to Play

1. Terry cut her paper into eight rectangles. Then she cut one of her rectangles into two halves. How many pieces would she have had if she had cut all eight of her rectangles into halves?

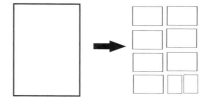

2. Chris cut his paper into eight rectangles. He put three of the rectangles into his mouth and chewed them. How many rectangles were left on the table?

3. Kelly raised her hand and asked for help cutting. Of all the kids in the class, Kelly demanded 40% of Carrie's time. How much time did that leave for the other kids? (All of Carrie's time is 100%. Subtract 40% from 100%.)

4. Suppose your piece of paper was in the shape of a hexagon. Could you make one straight cut and turn it into triangles?

hexagon

·······ANSWERS·······

1. If she cut all 8 rectangles in half, she would have 16
pieces.   8
$$\begin{array}{r} 8 \\ \times\ 2 \\ \hline 16 \end{array}$$

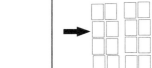

2. Start with 8.   Remove 3.   You have 5 left.

$$\begin{array}{r} 8 \\ -\ 3 \\ \hline 5 \end{array}$$

3.
$$\begin{array}{r} 100\% \\ -\ \ 40\% \\ \hline 60\% \end{array}$$
   Carrie would have 60% of her time
for her other students.

4. I can't see how to do it with one cut.
It looks like it would take at least 3 cuts.

I could make lots of triangles with 7 cuts!

Oops! I'm wrong. That's
not a triangle, but a pentagon.
(Pentagon = 5 sides)

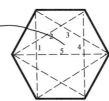

Chapter Twelve
Ducky or Honey

The arts and crafts was going as well as could be expected for five-year-olds. Kelly (who loved to raise her hand) had her left hand up in the air while she tried to cut the paper with her right hand.

Chris had five dry rectangles and three wet ones. Chris also took a taste of the honeycomb.

Terry had nine rectangles. She hadn't paid attention.

Lee, Dale, and Fred each had eight rectangles.

Carrie handed out crayons to the class and asked them to draw a bee on one of their rectangles.

Lee's drawing

Terry's drawing

Fred's drawing

Carrie explained to the class, "Now in the O.T., nobody did beekeeping. All the bees were wild."

Kelly raised her hand and asked, "What does O.T. stand for?"

Terry guessed, "Olden Times."

Carrie continued, "Some wild bees made their homes in holes in the ground. Some in cracks in rocks, and some in the bodies of dead animals."*

Everybody in the class said, "Yuck! Bees inside a dead animal."

It was the perfect moment for snack time. Carrie got out a bowl of honey and dipped graham crackers in it. She gave one to each kid.

Fred wasn't that hungry. He put his cracker on the table "for later."

Kelly raised her hand. She needed to go to the bathroom.

Chris got a spoon and started working on the bowl of honey. When Carrie saw this, all she could think of was Proverbs 25:16: "If you find honey, don't eat too much, or you'll throw up." Soon Chris needed to head to the bathroom.

He didn't raise his hand. He just ran.

---

* Carrie wasn't inventing this. She had carefully researched this. 1 Samuel 14:25; Deuteronomy 32:13; and Judges 14:8.

After Sunday school was over, Fred put the graham cracker with honey on it into the garbage can. He didn't want to put it into his pocket.

Then it hit him. Fred had been trying to think of how to have his arithmetic students learn the six facts for the four times table.

$$
\begin{array}{cccccc}
4 & 5 & 6 & 7 & 8 & 9 \\
\times 4 & \times 4 & \times 4 & \times 4 & \times 4 & \times 4 \\
\hline
16 & 20 & 24 & 28 & 32 & 36
\end{array}
$$

> Read these aloud!

On one of the blank pieces of paper Fred wrote:

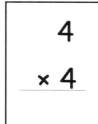

and on the back he wrote:

16

Fred had reinvented flash cards. He called them honey cards. He didn't know that you could go to King KITTENS and buy printed **Ducky** brand flash cards.

Handy Comparison Chart

vs.

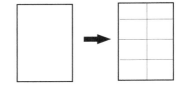

Fred's do-it-yourself Honey Cards

Cost	
Several dollars.	As cheap as a sheet of paper.

Convenience	
Already printed.	When you write it yourself, you learn more than when it is already printed for you.

Durability	
Will last as long as playing cards.	If you lose one card, you can make another.

Availability	
Drive to the store. Park. Find the cards. Stand in line. Pay for them and pay sales tax. Get back to the car. Drive home. Bring cards into the house. Open the package.	Cut paper into eight rectangles and start writing. If you happen to have some 3x5 index cards, you don't even have to do any cutting.

Your Turn to Play

It's Fred or **Ducky** time. The multiplication table is going to be yours today—from 2×2 up to 9×9.

Your choice:

Either . . .

A) Get your mother/father/butler/ maid/older brother/uncle/aunt to drive you to the store and buy the commercially made flash cards, or

B) Get out some paper and scissors (or some index cards) and make your own Fred's Honey Cards.

Making Fred's Honey Cards

Cut up the paper into eight rectangles.

On one side write the problem.

For example: 7
                   × 4

On the other side, write the answer: 28

Here are the eight facts for the two times table. Copy one fact on each of the eight rectangles.

2	3	4	5	6	7	8	9
×2	×2	×2	×2	×2	×2	×2	×2
4	6	8	10	12	14	16	18

We are going to need to cut up more rectangles. Cut up four more sheets. That will give us 32 more rectangles. $(4 \times 8 = 32)$

Here are the rest of the tables to copy.

3	4	5	6	7	8	9
×3	×3	×3	×3	×3	×3	×3
9	12	15	18	21	24	27

4	5	6	7	8	9
×4	×4	×4	×4	×4	×4
16	20	24	28	32	36

5	6	7	8	9	6	7	8	9
×5	×5	×5	×5	×5	×6	×6	×6	×6
25	30	35	40	45	36	42	48	54

7	8	9		8	9		9
×7	×7	×7		×8	×8		×9
49	56	63		64	72		81

You have four blank rectangles left over. Put these "upside downs" on them.

6	7	7	8
×9	×8	×9	×9
54	56	63	72

All done! ☺

Chapter Thirteen
Fred's New Hobby

Before you begin this chapter, you have a little work to do to <u>earn the right</u> to continue reading Fred's adventures. It will only take a minute or two.

Take the 21 honey cards for the two times table, the three times table, and the four times table.

Take each card and say (or guess) the answer and see if you got it right.

Fred helped Carrie straighten out the chairs after Sunday school was over. He put the phonebook back where he had found it.

Carrie headed off to the church service, and Fred went outside. He wasn't old enough for adult church yet.

Fred thought to himself, *I wonder how hard it would be to be a beekeeper? I could make a hive—that's what Carrie called it—and rent it out to bees. Then I could collect the rent, which in this case would be honey.*

The only thing that Fred didn't know about beekeeping was . . . everything! He didn't know how to make a hive, how to get bees to live in it, and how to get the honey out of the hive.

Fred ran back to his office. The walls of his office were lined with books. He had alphabetized all his books. He knew exactly where to look.

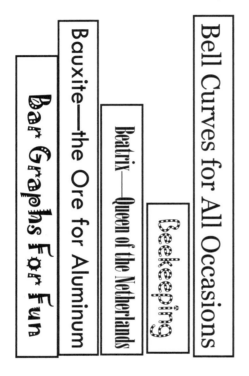

He had bought the book on bauxite in case he ever needed to make some aluminum foil "from scratch." In another part of his library he had a book on smelting iron to make steel.

He always liked the Netherlands. He could tell exactly how far the Netherlands is from Holland. Every well-educated adult knows the exact distance.

There it was! Prof. Eldwood's book on beekeeping.

Fred knew that the Latin word for bee is *apis*. So he knew that apiarist was someone who took care of bees. (A-pea-eh-rist. Accent the A.)

He took the book to his desk and started to read. He skipped the first chapter that described the long history of apiculture (keeping of bee colonies).

He read about how bees make honey. At the first sign of daylight, some of the bees head out to gather nectar from flowers. They can travel farther than you could walk in 20 or 30 minutes.[*]

They haul the nectar back to the hive and transfer it to bees that work inside the hive. Those bees stuff the nectar into the honeycomb cells.

hexagons

It's not yet honey. The bees have to go one step further. The nectar is too wet. To dry it out they fan their wings over the comb. When the water content is down to 17%, they cap each hexagonal cell with wax. It is now honey.

17% water means 17 parts of water out of a 100.

$$\begin{array}{r} 17\% \text{ water} \\ +\ \underline{83\%} \text{ not water} \\ 100\% \end{array}$$

---

[*] We use *farther* for actual physical distance and *further* for everything else.

Honey comes in many different colors and tastes depending on which plants the bees visit.

Fred trembled with excitement. He flipped to the chapter on creating an apiary (A-pea-airy—a place where you keep your bees).

There were two parts: (1) buying all the stuff and (2) doing all the work.

Fred figured that he would get all the supplies ordered first. Then he could spend the time while he was waiting for the stuff to arrive to learn all about apiculture.

Big Error!

Have you ever noticed how much easier it is to shop than to study?

What if Fred finds out that beekeeping isn't as much fun as he imagines it to be?

Fred didn't have a workshop so he decided that it would be easier to buy the hives than to build them. Four hives seemed to be a nice number.

A three-pound package of bees would cost around $90. It included one queen bee. Fred

didn't know anything about beekeeping. He
wondered why there wasn't a king bee.

Your Turn to Play

1. If one package of bees costs $90, how much would four packages cost?

2. If one hive costs $73, how much would four hives cost?

3. Then he would need a bee hat that had a net so he wouldn't get stung on the face. (Size extra-small)

He would need a bee hat for Kingie. (Size extra-extra-small)

He would need a smoke generator to quiet the bees when he was gathering the honey.

Fred's bee hat	52
Kingie's bee hat	47
Smoke generator	72
A hot knife to cut the wax off the comb	100
A hand crank extractor	300
Empty jars	25
Labels for the jars	+ 6

How much will all these items cost?

·······**ANSWERS**·······

1.   90
     ×   4
   ─────
     360   Four packages of bees will cost $360.

2. ^{1}73
     ×   4
   ─────
     292   Four hives will cost $292.

3.   Fred's bee hat ^{2}52
            Kingie's bee hat   47
            Smoke generator $_2$ 72
A hot knife to cut the wax off the comb   100
A hand crank extractor   300
            Empty jars   25
            Labels for the jars   +   6
                                               ─────
                                                 602

All these materials will cost $602.

If we add in the cost of the bees (from question 1) and the cost of the hives (from question 2), we have . . .

        360
        292
     +   602   His total cost to set up an
     ─────   apiary will be $1,254.
       1254

Chapter Fourteen
Starting a Business

Before you begin this chapter, you have a little work to do to <u>earn the right</u> to continue reading Fred's adventures. It will only take a minute or two.

Take the 21 honey cards for the two times table, the three times table, and the four times table.

Take each card and say (or guess) the answer and see if you got it right.

Make two piles: the ones you got right and the ones you missed. Go through the missed pile a second time.

Fred would have to borrow $1,254 (one thousand, two hundred fifty-four dollars) from his doll Kingie. Fred didn't have any money because C.C. Coalback had stolen it from him.

Fred told Kingie, "I'm thinking of starting a business."

Kingie was delighted. "You know that people who start a business have a better chance of making big money than those who work for a salary. You should also note that many small businesses go broke."

Kingie, businessman

Fred had the perfect situation. He had a salary, and he could start a business. Then if the business didn't succeed, he could still live on his salary.

Kingie made and sold his oil paintings. He was the first successful businessman that Fred had known.

Kingie pulled out his famous . . .

Checklist for
Starting a Business

1. **Why are you choosing this particular business?**
2. **Does it make financial sense?**
3. **Are there any obvious drawbacks?**

Kingie started with question number one and asked, "What business are you thinking of, and why have you chosen it?"

Fred was delighted to talk about his fascination. "I am going to be an apiarist. I think it will be sooooo neat. I'm all excited. I first learned about it in Sunday school this morning."

Kingie had no idea what an apiarist was. He thought it might be someone who grows pears.

There are two bad answers to the question of why a particular business is being chosen.

Bad answer #1: I'm going into this business to make a lot of money. The reason this is a bad answer is that running a business takes a lot more time and effort than holding down a job. Twelve-hour work days are not uncommon for business owners. You should not be opening a pet store if you don't really love animals. You should not be a doctor if you don't like being around the sick and dying.

Bad answer #2: I just heard about this business and I think it's cool. The reason this is a bad answer is that starting a business is a major commitment of both study time and money. You don't decide to open a bowling alley because you went bowling last week for the first time in your life, and you can't get bowling balls out of your brain.

For most businesses, it takes 1,000 hours of study before you start the business. One thousand hours equals four hours of study per day for a year. (Five days per week and 50 weeks per year.)

To be a master in most businesses, 5,000 hours of study is the usual minimum.

Kingie rated Fred's answer to **1. Why are you choosing this particular business?** with a big ☺.

The second question Kingie asked Fred was whether being an apiarist made financial sense. Here is one place where knowing some math can spell the difference between success and failure.

"What are you going to be selling?" Kingie asked.

"Delicious jars of honey," Fred said. "I need to borrow $1,254 from you for all the supplies. I figure that I will make a profit of $3 for each jar I sell."

When you want to see if a particular investment makes financial sense, you have to do the arithmetic. In the language of the business world, Kingie needed to find out whether Fred's proposed business would **pencil out**.

How many jars would Fred have to sell in order to get back his investment of $1,254? He makes a profit of $3 per jar.

Do we add, subtract, multiply, or divide? That's always **the big question in arithmetic**.

> **If you don't know whether to add, subtract, multiply or divide, first restate the problem with really simple numbers.**

With really simple numbers—suppose he would make a profit of $3 per jar, and he wanted to get back his investment of $6. ← really simple number

We know he would have to sell 2 jars. How did we get the 2? We divided 3 into 6.

This simple procedure we will use over and over again. (1) Use some simple numbers. (2) The answer pops into your head. (3) At that point see whether you added, subtracted, multiplied, or divided.

Your Turn to Play

1. Pencil out Fred's business. How many jars would Fred have to sell in order to make $1,254. He makes a profit of $3 per jar.

If he wanted to make $6, we would have $3\overline{)6}^{\,2}$

2. Would you add, subtract, multiply, or divide in this problem? (You are not being asked to solve the problem!)

If each jar of honey weighed 482 grams, how many grams would 57 jars weigh?

3. Would you add, subtract, multiply, or divide in this problem? (You are not being asked to solve the problem!)

If 170 bees weigh 850 grams, how much does each bee weigh?

. . . . **ANSWERS** . . . .

1. We know that we need to divide.

$$\begin{array}{r} 418 \\ 3{\overline{\smash{\big)}\,1254}} \\ \underline{12} \\ 5 \\ \underline{3} \\ 24 \\ \underline{24} \\ 0 \end{array}$$

> Fred would have to
> sell 418 jars in order
> to make $1,254.

2. We substitute some simple numbers: *If each jar of honey weighed 3 grams, how many grams would 2 jars weigh?*

We know the answer would be 6 grams. Now we look to see how did we got the 6. We multiplied.

3. We substitute some simple numbers: *If 2 bees weigh 6 grams, how much does each bee weigh?*

We know the answer is 3 grams. Now we look to see how we got the 3 grams. We divided.

If we wanted to solve the original problem: *If 170 bees weigh 850 grams, how much does each bee weigh?* we would divide 170 into 850. (You haven't had the seven times table yet, so we can't actually do the division.)

Chapter Fifteen
Unstoppable Fred

Kingie shook his head. Fred was going to have to sell 418 jars of honey just to get back his original investment. He would have to sell 418 jars just to break even. When he sold his 419th jar, he would make three bucks.

Here's what 418 jars looks like:

$$400 + 10 + 8$$
$$= 418$$

There is no way that Fred could sell that many jars. Kingie rated Fred's answer to **2. Does it make financial sense?** with a big ☹.

At this point Kingie told Fred that he had flunked the first two questions on the checklist for starting a business: he was choosing the business for the wrong reason, and the business didn't pencil out.

Did this advice stop five-year-old Fred? He was busy drawing up a sign to put on the front door of his office:

"What are those things on your sign?" Kingie asked. "And what's an apiarist?"

"I'm going to make my office into an apiary—a place you keep bees," Fred answered. "I'm going to start out with only four hives."

If dolls could faint, Kingie would have passed out.

It was clearly time for the third item on Kingie's checklist for starting a business: **3. Are there any obvious drawbacks?**

Question: What's an obvious drawback to becoming a typewriter manufacturer?

Answer: Almost no one uses typewriters nowadays.

Question: What's a drawback to farming marijuana?

Answer: It's illegal.

Question: What's a drawback to selling 50-pound dumbbells to nursery schools?

Answer: If you can't figure that out, I won't tell you.

Kingie asked the third question from the checklist: "Do you see any possible reasons why having four hives in your office might not work?"

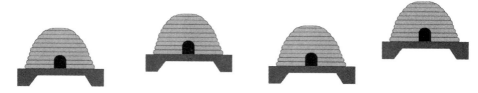

Fred was all excited about beekeeping. He said, "It would be a lot of fun."

Kingie was going to have a tough time with this kid. "I didn't ask if you *want* to do it. I asked if there were any possible drawbacks."

Fred started singing to himself, "♪You got to accentuate the positive ♫ eliminate the negative." It was part of a song that his great-grandfather might have sung. Fred wasn't in the mood to be bothered with the REALITY of beekeeping in a math office.

small essay

Reality

You may remember what Sir Freddie Laker said on February 4, 1982: "I'm flying high and couldn't be more confident about the future." On February 7th, his Laker Airways business collapsed.

end of small essay

People ignore reality all the time . . .

He washed his hair in gasoline and then lit up a cigarette.

She drove her motorcycle on icy twisting roads at night.

He ate a big bowl of honey.

She only knew him for a week . . . and married him.

He bought a pet alligator. His kids loved to play with it.

but that doesn't mean that reality will ignore them.

Kingie had to make his own private list of the drawbacks. Fred wasn't in the mood to listen right now. Hopefully, before Fred ordered all the beekeeping supplies, Kingie could show him the list. Kingie has nice handwriting. He's an artist.

Drawbacks

1. There isn't enough room in the office for four hives.

2. *Some people are allergic to bee stings. What if some student visits you, gets stung, and ends up in the hospital?*

3. *The bees have to get outside to gather the nectar. That means you will have to leave the window open. This is February in Kansas. We'll freeze!*

Your Turn to Play

1. Fred would have to sell 418 jars of honey just to break even. Is this a cardinal or an ordinal number?

2. When he sold his 419th jar he would start to make money. Is this a cardinal or an ordinal number?

3. $2\frac{1}{2} + 1\frac{1}{2} = ?$

4. In order to get the honey out of the hive, you have to pull out the panels that hold the honeycomb.

Sometimes the bees are in a happy mood and won't mind that you are taking apart their hive and taking their honey.

Sometimes they are not in a happy mood.

Take my honey. I've got lots.

Do you want to get stung?

That's why Fred was going to buy a $72 smoke generator. Sticking smoke in the hive makes the bees dazed and less likely to sting. If Fred paid for the smoke generator in 3 equal payments, how much would each payment be?

. . . . ANSWERS . . . .

1. Four hundred eighteen is a cardinal number. Cardinal numbers are used to describe the number of members of a set.

2. Four hundred nineteenth is an ordinal number. Ordinal numbers are first, second, third, fourth. . . .

3. 2 ½ + 1 ½
   = 4

4. We want 3 equal payments to buy a $72 item. Do we add, subtract, multiply, or divide?

   This is the big question in arithmetic.

> **If you don't know whether to add, subtract, multiply or divide, first restate the problem with really simple numbers.**

With really simple numbers: We want 3 equal payments to buy a $12 item. Each payment would be $4. How did we get the 4? We divided 3 into 12.

   So, in the original problem, we need to divide 3 into 72.

$$\begin{array}{r} 24 \\ 3\overline{)\,72} \\ 6 \\ \hline 12 \\ 12 \\ \hline \end{array}$$

Three payments of $24

Chapter Sixteen
Fred Stopped

Do you have a minute?

Before you begin this chapter, you have a little work to do to <u>earn the right</u> to continue reading Fred's adventures.

Here is the Official Procedure for honey cards. (This is something we have not mentioned before.)

Take each card and say (or guess) the answer and see if you got it right. Put the ones you got right in one pile and the ones you missed in another pile.

Now pick up the missed pile *and repeat* until all the cards end up in the I-got-it-right pile.

In order to start a successful business, *all three items* on Kingie's checklist must be answered with a ☺.

In Fred's dream of having an apiary, none of them received a ☺.

1. He was choosing the business that he really didn't know much about. ☹

2. It didn't pencil out. He would lose money. ☹

3. There were major drawbacks. ☹

But Fred was unstoppable. He wanted to be a beekeeper. He needed to be a beekeeper.

It was something that he always wanted to do.*

Question: Do five-year-olds ever get so wrapped up in an idea that they stubbornly insist on charging ahead even though there is every indication that the results will be disastrous?

Answer: Yup.

Question: Do teenagers ever get so wrapped up in an idea that they stubbornly insist on charging ahead even though there is every indication that the results will be disastrous?

Answer: Yup.

Question: Do twenty-five-year-olds ever get so wrapped up in an idea that they stubbornly insist on charging ahead even though there is every indication that the results will be disastrous?

Answer: Yup.

Question: Do forty-five-year-olds ever get so wrapped up in an idea that they stubbornly insist on charging ahead even though there is every indication that the results will be disastrous?

Answer: Yup.

Question: Do really old people (over 50) ever get so wrapped up in an idea that they stubbornly insist on charging ahead even though there is every indication that the results will be disastrous?

Answer: Yup.

Question: Are people over 50 really old?

Answer: Only in the minds of people under 30.

---

* From Fred's point of view, he had always wanted to be an apiarist. In reality, it was now eleven o'clock. He had first learned about beekeeping less than two hours ago.

Fred changed into his shopping clothes. He headed off to the largest store in town: King KITTENS.

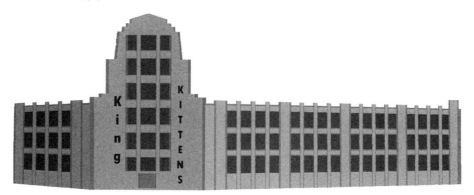

It had three main floors, and each floor was seven acres. Somewhere in those 21 acres of shopping space he would find the $1,254 of supplies in order to become an apiarist.

Everything except . . . the $1,254. Fred had no money of his own. Kingie wasn't about to lend him a dime.

At the main entrance to King KITTENS was the famous KKKKK (King KITTENS Kredit Kard Korner).

How Kute! Fred thought to himself. I can get a credit card. There were three different cards available.

Disaster Card	Tower of Pisa	American Depress
Easy Credit!	Easy Credit!	Easy Credit!

The people at all three credit card stations tried to hand Fred an application. One of them was a very pretty woman. The second was a man in a business suit. Fred took the application from the clown with the funny hat. He thought that that company would be most kid-friendly.

He started to fill out the credit card application.

NAME: _Fred Gauss_

ADDRESS: _room 314, Math building, KITTENS_

LENGTH OF TIME AT CURRENT ADDRESS: _4 years, 6 months_

And then he looked at the last line:

I, THE UNDERSIGNED, DECLARE THAT THE ABOVE STATEMENTS ARE TRUE AND THAT I AM AT LEAST I 8 YEARS OLD.

YOUR SIGNATURE:_____

Fred was doomed.

$$\begin{array}{r} 18 \\ -\ 5 \\ \hline 13 \end{array}$$

He had to wait 13 years before he could head off to King KITTENS and buy things he couldn't pay for with his savings.

Your Turn to Play

1. Fred was already at King KITTENS, so he decided to go inside and just look around.

He passed the food court that was located on the main floor. At the KKKKK (King KITTENS Krazy Kone Koncession) they were offering half scoops of ice cream.

A boy asked for five half scoops.

½ + ½ + ½ + ½ + ½ = ?

2. He remembered that he didn't have a sleeping bag. Fred had thrown his old one out of the window, and it was caught in a tree.

Right next to the food court was SSSSS (Sam's Slumber Shop for the Super Short).* Fred looked in the window and saw the sign: We Sell Sleeping Bags by the Inch. $4 for each inch.

Fred is 36 inches tall. How much would his sleeping bag cost?

---

* TTTTT (Timothy Tom's Tailoring for the Terribly Tall) was right next to SSSSS. Everyone called him Tim Tom, the tailor.

. . . . ANSWERS . . . .

1. ½ + ½ + ½ + ½ + ½ = 2 ½

2. Sleeping bags are $4 per inch. Fred is 36 inches tall. Do we multiply or divide?

> If you don't know whether to multiply or divide, first restate the problem with really simple numbers.

Suppose sleeping bags are $2 per inch and Fred is 3 inches tall. Then his sleeping bag would cost $6. How did we get the $6? We multiplied.

So, in the original problem, we want to multiply 4 times 36.

$$\begin{array}{r} 2 \\ 36 \\ \times\ 4 \\ \hline 144 \end{array}$$

His sleeping bag would cost $144.

That may seem like a lot of money for a sleeping bag, but note that these are custom-made. Six-foot sleeping bags made in factories cost less because they are mass-produced. It would be cheaper for Fred to buy a six-foot sleeping bag and just use the top half.

Chapter Seventeen
Fishing

Did you notice that in the previous Your Turn to Play we started assuming that you know the four times table?

Before you begin this chapter, you have a little work to do to <u>earn the right</u> to continue reading Fred's adventures.

Take each card and say (or guess) the answer and see if you got it right. Put the ones you got right in one pile and the ones you missed in another pile.

Now pick up the missed pile and repeat until all the cards end up in the I-got-it-right pile.

Fred headed back to his office. It wasn't much fun looking at the new sleeping bags if he couldn't afford one.

He stared out the window at his old sleeping bag. He had really made a mistake in throwing it out of the window instead of carrying it down to the laundry on the first floor.

And trying to act like a cowboy and lasso his sleeping bag was another mistake. Sometimes it is very difficult to look back at the mistakes we have made in our lives.

For years Fred would have to look at that sleeping bag in the tree. It was like looking at a tattoo that you wish you had never gotten.

It was almost noon. Fred wasn't very hungry. He decided to read for a while.

★   ★   ★

Darlene and Joe were taking a walk on the KITTENS campus after church. They are two students in Fred's arithmetic class that meets at 8 a.m.

Darlene liked walking with Joe. When she read her bridal magazines, she always thought of him.

Joe liked walking with Darlene because he could tell her all about his fishing adventures. She was one of the few people who would listen to him.

One of his favorite stories was about the time he was fishing in the Cheney Reservoir near Wichita. It was on the seventh of July. He remembered that date since it is sometimes written as 7/7. July is the seventh month of the year.

"I caught a million fish on that day," Joe bragged. "You should have seen how many I caught."

Darlene didn't remind Joe that she had been there with Joe. She had cleaned the 37 fish that he caught.

$$1,000,000 \quad \hookleftarrow \text{Joe's story}$$
$$\underline{- \qquad\quad 37} \quad \hookleftarrow \text{Actual number}$$
$$999,963 \quad \hookleftarrow \text{The difference between}$$
$$\text{the truth and his story}$$

He continued, "The biggest fish I caught that day was 180 inches."

To convert inches to feet, you divide by 12. How do we know that we should *divide* instead of add, subtract, or multiply?

You restate the problem with really simple numbers. If something is 24 inches, then it would be 2 feet. You went from 24" to 2' by dividing.

$$12\overline{)180} \qquad\qquad \text{Division by a two-digit number}$$

$$\begin{array}{r} 1 \\ 12\overline{)180} \end{array} \qquad \text{12 goes into 18 once.}$$

$$\begin{array}{r} 1 \\ 12\overline{)180} \\ \underline{12} \end{array} \qquad \text{1 times 12 is 12.}$$

$$\begin{array}{r} 1 \\ 12\overline{)180} \\ \underline{12} \\ 60 \end{array} \qquad \text{Subtract and bring down the 0.}$$

$$
\begin{array}{r}
15 \\
12\overline{)180} \\
\underline{12} \\
60
\end{array}
$$

12 goes into 60 five times.

$$
\begin{array}{r}
15 \\
12\overline{)180} \\
\underline{12} \\
60 \\
\underline{60} \\
0
\end{array}
$$

Five times 12 is 60.

His 180-inch fish is 15-feet long. Darlene never mentioned to Joe that a 15-foot fish would never have fit in his eight-foot boat.

Actually, the fish that Joe had caught was only 18 inches, not 180 inches. To convert 18 inches into feet, we divide by 12.

$$
\begin{array}{r}
1 \text{ R } 6 \\
12\overline{)18} \\
\underline{12} \\
6
\end{array}
$$

Our answer is 1 foot with a remainder of 6.
Joe's fish was 1 foot, 6 inches long.

Your Turn to Play

1. Dividing by two-digit numbers such as 12)‾‾‾‾ or 37)‾‾‾‾ is a little harder than dividing by single-digit numbers.

   Sometimes you have to *guess* how many times the **divisor** goes into the number.

12)‾‾‾‾

For example, if you had   12)‾64

you might guess too much
$$\begin{array}{r} 6 \\ 12\overline{)64} \\ 72 \end{array}$$

> Can't subtract.
> Bottom number too big.

or you might guess too little
$$\begin{array}{r} 4 \\ 12\overline{)64} \\ 48 \\ \hline 16 \end{array}$$

> $16 > 12$
> When you subtract, the answer should be less than the divisor.

Your turn to play.
Divide 12 into 64.

2.   24)‾75

······**ANSWERS**·······

1.
$$12\overline{)64}$$
$$\begin{array}{r} 5 \ \ R\,4 \\ \hline 60 \\ \hline 4 \end{array}$$

2.
$$24\overline{)75}$$
$$\begin{array}{r} 3 \ \ R\,3 \\ \hline 72 \\ \hline 3 \end{array}$$

In the first problem above, you had to multiply 5 times 12. You might have wondered how you were expected to multiply by 5 when we haven't gotten to the 5 times table yet.

Easy. We have done (and you have been practicing with your honey cards) the 2 times table. You know that 2 × 5 is 10.

So you already know that 5 × 2 is 10.

And 5 × 3 = 15. And 5 × 4 = 20.

Chapter Eighteen
Free Stuff

Add the five times table cards to your honey cards. There are five new facts to practice: $5 \times 5 = 25$ $5 \times 6 = 30$ $5 \times 7 = 35$ $5 \times 8 = 40$ $5 \times 9 = 45$

Before you begin this chapter, you have a little work to do to <u>earn the right</u> to continue reading Fred's adventures.

Take each card and say (or guess) the answer and see if you got it right. Put the ones you got right in one pile and the ones you missed in another pile.

Now pick up the missed pile and repeat until all the cards end up in the I-got-it-right pile.

Zeros meant nothing to Joe. He had caught an 18-inch fish and called it 180 inches. Last week, when he told Darlene the story of his July 7th (7/7) fishing trip, he said that the fish was 1,800 inches.

$$\begin{array}{r} 150 \\ 12\overline{)1800} \\ \underline{12} \\ 60 \\ \underline{60} \\ 00 \\ \underline{00} \end{array}$$

1,800 inches is equal to 150 feet.

A football field is 100-yards long. There are three feet in a yard. To convert 100 yards into feet, do we multiply by three or divide by three?

> **If you don't know whether to multiply or divide, first restate the problem with really simple numbers.**

With really simple numbers: We know that 2 yards is 6 feet. We multiplied.

So 100 yards equals 3 × 100 feet. 300 feet

So if Joe's fish were 150-feet long, then two of them would be the length of a football field.

$$\begin{array}{r} 150 \\ \times\ 2 \\ \hline 300 \end{array}$$

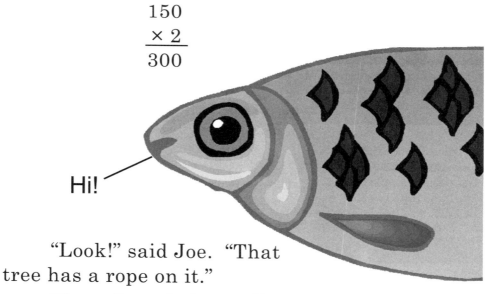

Hi!

"Look!" said Joe. "That tree has a rope on it."

First, Joe tried to pull the rope down. He figured that he could have a free rope. Joe liked free stuff. It cost zero dollars.

Second, when the rope wouldn't come down, Joe swung on the rope like Tarzan.

Third, Joe climbed up the rope.

"Lookie!" he exclaimed. "I found a big pillow in the tree. It's free." He threw it out of the tree.

Joe untied the rope and threw that down also. He now had a free rope.

Then he realized that without the rope he couldn't get out of the tree. Joe was stuck.

He panicked.* He started screaming, "Help! Help! Fire! Earthquake! Hurricane!" He was so panicked that he didn't know what he was saying.

It would have been a lot more sensible just to tell Darlene that he was stuck and ask her to go get some help.

Darlene said, "Hold on. I'll go and get some help."

Joe said, "I'm holding on."

---

* *Panic* is a verb. The present tense is *panic*. When there is no pizza in the house, I panic. The past tense is *panicked*. Yesterday, when my favorite pizza store went out of business, I panicked. Do you see the *k* in the past tense?

I like to shellac my art work. Yesterday, I shellacked three drawings.

You can mimic me as I frolic in the world of math. Yesterday, I mimicked a garbage disposal as I frolicked in a large combination pizza.

What Darlene meant by *hold on* was to wait. What Joe understood by *hold on* was to not let go.

Darlene said that she was *going* to get some help. The word *go* in my dictionary has 84 different meanings.

Math is much easier than English.

In algebra, 7xyz + 8xyz always equals 15xyz.

Darlene walked up the two flights of stairs to Fred's office. Fred was reading his Kleene's *Introduction to Metamathematics.* Fred had just gotten to the method "for enumerating the ordered pairs of members of an enumerable set" (page 5) when Darlene came in and said, "Joe's stuck in the tree outside your window. What should we do?"

Fred looked out his window. It seemed almost like magic.* It looked like Fred's sleeping bag had turned into Joe. That was quite a trick.

Fred didn't ask how Joe got stuck in the tree. That kind of thing happened to Joe a lot.

_____

* Magic only means things that we don't yet understand.

Fred headed to the janitor's office. His office wasn't hard to find. He had a big flashing, neon sign in front of his office.

SAMUEL P. WISTROM

Educational Facility Math Building
Chief Inspector/Planner/Remediator
for offices 225–324

This big sign meant that he was the janitor for half of the Math building.

Your Turn to Play

1. Let's do some algebra: $6wx + 9wx = ?$

$22x + 33x = ?$

$88y - 8y = ?$

2. Darlene walked up two flights of stairs to get to Fred's office. The first flight had 37 steps and the second had 36 steps. How many steps did she climb?

3. If it took 4 calories for each step, how many calories did Darlene burn walking up those two flights?

4. Darlene walked down the hallway (past the four vending machines on one side and five on the other) in order to get to Fred's office. The hallway is 492 inches long. How many feet is that?

·······ANSWERS·······

1.   6wx + 9wx = 15wx

       22x + 33x = 55x

       88y − 8y = 80y

2. $\overset{1}{37}$
   + 36
     73

     She walked up 73 steps.

3. Four calories per step. Seventy-three steps. Do we add, subtract, multiply, or divide? We restate the problem with really simple numbers: Suppose it was 2 calories per step and 3 steps. She would have burned 6 calories. We multiplied 3 times 2.

        4
     × 73   is funny looking. Since 4 × 73 is the same

          73 × 4, we'll rewrite the problem as

      73
    × 4
    292   She burned 292 calories.

4. To change 492 inches into feet, we divide by 12.

$$\begin{array}{r} 41 \\ 12\overline{)492} \\ \underline{48} \\ 12 \\ \underline{12} \end{array}$$
      The hallway is 41 feet long.

Chapter Nineteen
Getting a Ladder

Fred didn't go to Sam when it was just his sleeping bag in the tree. But when there was a human caught in the tree, it was a different matter.

As Fred and Darlene approached Sam's door they could hear the noon news that Sam was watching on television:

The weather will be the same as it has been for the last three months. It will be cold.

The stock market remains jittery.

There is trouble in the Middle East.

Computers are getting faster.

The government schools and the post office need more money.

In other words, there was no new news. Sam was just wasting time.

Fred knocked on the door. Sam couldn't hear him. Sam's television was loud. Years of going to rock concerts had made him partially deaf.

They walked into his office and stood in front of the television set. Sam didn't notice. He was asleep. They turned off the TV and gently woke him.

"Sam, we need your help," Fred said. "Joe's in trouble."

"What's it this time?" Sam asked. "Last week he got his foot stuck in a paint can. Before that he accidentally locked himself in a closet."

"Joe climbed up the big tree outside the Math building," Darlene explained, "and he can't get down."

She omitted the part about Joe using a rope to climb up the tree and then throwing the rope down. She didn't want people to think that Joe was stupid.

"I guess I'll need my big ladder," Sam said as he got out of his chair.

The three of them headed outside to the back of the Math building where Sam had stored his largest ladder.

There was a big sign on the ladder: **JOE! STAY OFF!**

Joe was the only university student that might climb that ladder for fun and get hurt.

Fred noticed that the **slope** of that ladder was 8.

Here's how to find the slope. You divide the rise (40') by the run (5').

$$5 \overline{)\, 40 \,}^{\,8}$$

40'

5'

The slope of a line is defined as the rise divided by the run.

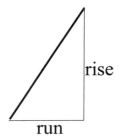

$$\overline{\text{run})\,\text{rise}}$$

slope = rise ÷ run

÷ means "divided by."

When you get to fractions, we can write this as

$$\text{slope} = \frac{\text{rise}}{\text{run}}$$

Preview of Coming Attractions:

In *Life of Fred: Beginning Algebra*, we'll give six ways to define slope. (It's *that* important.)

In *Life of Fred: Trig*, we will invent the tangent function which will turn angles automatically into slope.

For example, if you know that the angle is 76 degrees (you will learn about degrees in *Life of Fred: Geometry*), then on a scientific calculator you press ⁷⁶ and then hit the t an button and find that the slope is just a little bigger than ⁴.

In *Life of Fred: Calculus,* we get really weird. Chapter 4 of that book is simply entitled "Slope."

Given a point on a curve,

we can find the slope of the tangent line to that point.

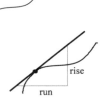

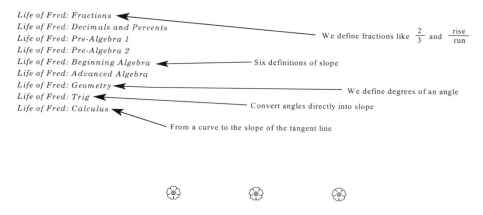

Sam and Darlene carried the ladder over to the tree. Sam took off the sign that read **JOE! STAY OFF!**

Joe hopped onto the ladder and climbed down out of the tree.

"Lookie!" Joe exclaimed. "I got some free rope."

Fred explained to Joe that the rope had been borrowed from the janitor's closet on the third floor hallway. It needed to be returned.

"But I've got a free big pillow," Joe continued.

They broke the bad news to Joe that the "big pillow" was really Fred's sleeping bag. Joe turned it over to Fred.

Fred took his sleeping bag to the laundry room to dry it. Darlene took Joe out for lunch. Sam headed back to catch the rest of the noon news.

Index

Index

To have your questions about
the Life of Fred series answered
or
to order
visit
FredGauss.com